The Comet Handbook

Garry Stasiuk and Dwight Gruber

My Name Is Halley

Imagine that you are a comet. This is your story.

Some 15 billion years ago, in a manner still not clearly understood, the universe expanded into being, forming the elements hydrogen and helium. Over billions and billions of years stars coalesced from this material, lived through their life cycles, and exploded, spewing into space new elements—carbon, nitrogen, oxygen, silicon, iron. These became raw material for the universe as we know it today. About five billion years ago, some event—a supernova, or an explosion in the core of the Milky Way galaxy—sent a shock wave through a nebulous cloud of this material, and started the collapse of this nebula which would eventually form our solar system.

When your parent star, our Sun, first formed, it was twice as large as it is now, and quite unstable. Matter and energy flowed away from the Sun at an enormous rate. This solar wind was so strong that it swept away gas and dust from the interior of the solar nebula, literally blowing them out of the solar system. The solar wind was strong enough to affect heavier collections of matter, to change their orbits so that they fell into nearby planets, into the Sun, or were hurled to the outer reaches of the solar system. It was in this chaos that you had your beginnings, and why you find yourself so far from the Sun that it appears only as the brightest star in the sky.

You are now in a vast cloud thirty to 100,000 times farther from the Sun than is Earth, a vast deep-freeze where the temperature hovers near absolute zero. Gases and dust blown out by the super-solar wind have coated you, and as you slowly rotate you realize you belong to a collection of about 100 billion similar chunks of ice and dust, debris ranging in size from microscopic to continental, all orbiting the distant, dim Sun.

The gravitational force between yourself and the Sun is tenuous. It is as though you are perched precariously at the edge of a cliff—a small push, and gravity will make you fall toward the Sun; or a small nudge, and the tenuous gravitational bond is broken and you fly out, lost among the vast distances between the stars, forever.

You are pushed!

What exactly pushed you is not known. Perhaps it was a distant, passing star; or perhaps it was a fortuitous alignment of the large outer planets; or perhaps it was a combination of both. Whatever it was, it happened. You were pushed toward the Sun. Your journey and adventure have begun.

You are quite small, perhaps four kilometers (about three miles) in diameter. And you are quite fragile, a frozen ball of snowflake-like ices and dust. It will take you several million years to fall past the nine planets and the debris called the asteroids. However, even at the beginning of your fall toward the bottom of the solar cliff, you feel the increase in solar radiation and the solar wind. And, as you fall closer to the Sun, you begin to change.

The ices you are made of are very volatile. The Sun's radiation and the solar wind transforms a bit of your surface from a solid to a wispy gas, and soon your solid nucleus is surrounded by your atmosphere, or coma. The solar wind now strikes and reacts with the molecules and atoms in your coma as well as your surface. The molecules are changed and carried away from you, and from the Sun. You are beginning to develop a tail, but it will not become noticeable until you cross the orbit of Mars.

When you began your fall, your orbit would have brought you back out of the solar system, back to your original home far from the Sun. But life is never simple, and a combination of circumstances over several million years changed your destiny. Pockets of volatile gas explode with a rocket-like action that slows you down or speeds you up, and changes your rotation. The giant planets tug or push at you, speeding you up or slowing you down. These and other events dictate that you'll never return home. Instead of orbiting the Sun once every several million years, you will now orbit, on the average, in only a little over seventy-six years.

Out at the fringe of the solar system you barely felt the effects of the Sun's radiation and solar wind. But now, at the orbit of Mars, their effect has become noticeable. Your coma grows to enormous proportions, as does your tail. And not one, but two tails grow. Like Mary's little lamb, you leave a bit of your tail behind you. Pointing directly away from both you and the Sun is a bluish-colored tail composed of ionized gases. Curving away from you is a faint yellowish-colored tail composed of dust.

On this trip past the Sun you are not noticed on Earth until a few days after your perihelion passage. A Babylonian priest/astronomer standing atop his ziggurat observatory notices a strange, sword-like, sweeping glow in the eastern sky before sunrise. Just before the Sun appears you are fully revealed in all your glory. To the observer you are an apparition, an unwelcome sign in an uneasy time. It is 1404 B.C., and your true nature is completely misunderstood.

Thirty centuries will pass before even a part of that nature is deciphered. In 1682 A.D. an Englishman, Edmund Halley, notices that your path in the sky seems to be the same as the paths taken by comets seen in 1531 and 1607. In 1695, using the material published in Isaac Newton's *Principia Mathematica*, Halley's computations predict that this bright comet will return in 1758. An amateur astronomer sights you on December 25, 1758.

Now you know who you are. Your name is Halley's comet, a Great Comet that returns to be visible from Earth about every seventy-five years. Your name is a memorial to the spirit of curiosity, to the desire to know all that can be known about the mysteries of the universe. In 1986 your apparition will be the beautiful, majestic display of nature that for at least forty centuries people of Earth have experienced and shared. This time they wait for your return with telescope and spacecraft to unlock more of the truth of your nature. But, these, too, are troubled times. And you wonder, will there be anyone to observe you when you return in 2061?

The History Of The World According To Comets

To the Greeks, *astron kometes*; to the Romans, *aster cometas*: the long-haired stars. To the Egyptians, the heiroglyph for the sky-goddess Nut, with long hair, roughly translated as "the woman with the long flowing hair."

Observations of Great Comets appear throughout written history, and from these writings emerges a common theme—if something bad happens, blame it on a comet! Even our word, disaster, derives from Latin: dis (evil or bad) aster (star). These "evil stars" have been blamed for war, famine, flood, pestilence, plague, poison in the atmosphere, death of rulers, higher taxes, and emission of dangerous rays!

Some of the things people have believed about comets are at once silly and tragic; others, interesting and profound.

In Babylon, a scribe etched forever into clay that in 1140 B.C.,

> . . . a comet arose whose body was bright like the day, while from its luminous body a tail extended like the sting of a scorpion.

Homer, writing in the *Iliad* sometime between 1200 and 800 B.C., described Achilles's helmet:

> Like the red star, that from this blazing hair
> Shakes down disease, pestilence, and war.

The Parsees of Persia, a Zoroastrian sect around 700 B.C. which believed in demons and fairies, claimed that ". . . comets are wicked fairies, which bring disease and death . . .". At the same time, the Chaldeans believed that comets were like the planets, and they traveled in circles around the Sun. Two hundred or so years later Pythagoras believed that planets and comets were at different distances from the Earth, and had their own separate paths.

Aristotle, after witnessing the Great Comet of 371 B.C., wrote,

> Comets are an exhalation of the Earth, an atmospheric phenomenon produced by gases burning in the 'region of fire' above the surface of the Earth.

> On the departure of the expedition of Timolean from Corinth to Sicily, the gods announced his success and future greatness by an extraordinary prodigy. A burning torch appeared in the heavens for an entire night, and went before the fleet to Sicily.

This was the meaning of a comet seen in 344 B.C. according to Diodorus Siculus.

Comets were often thought of as swords or knives, as in the Bible, 1 Chronicles 21:16, written between 332 and 167 B.C.:

> And David lifted up his eyes, and saw the angel of the Lord stand between the Earth and Heaven, having a drawn sword in his hand stretched out over Jerusalem. . .

Great Comets often were thought to concern the deaths of great leaders:

> The common people supposed that the comet indicated the admission of the soul of Julius Caesar into the ranks of the immortal gods,

wrote Suetonius, commenting on the Great Comet of 43 B.C.

Atilla the Hun's death was said to be foretold by comets in the years 453 and 455 A.D., even though Atilla died in 453!

> Thou art come! A matter of lamentation to many a mother art thou come; I have seen thee long since; but now I behold thee much more terrible, threatening to hurl destruction on this country,

wrote the Earl of Malmsbury, forecasting the fall of King Harold of England because of the appearance of Halley's comet in 1066 A.D. Tennyson, in his play *Harold*, called it "Yon grimly-glaring, treble brandished scourge of England."

King Harold died in the Battle of Hastings in 1066. Halley's comet is pictured in the Bayeux Tapestry, woven a few years after the battle to commemorate the Norman's victory. But the first realistic portrait of a comet was painted by Giotto de Bordone, in the Arena Chapel fresco in Padua. The painting of the comet represented the star of Bethlehem. It is believed that Giotto based the painting on his own observation of Halley's comet during its apparition of 1301.

The Aztecs believed that the appearance of the Great Comets of 1409 and 1499 were signs that their great god Quetzelcoatl was about to return.

Ambroise Pare, writing of a comet seen in 1528, said,

> This comet was so horrible, so frightful, and it produced such great terror, that some died of fear and others fell sick.

It is now believed that what Pare called a comet was actually a very bright and unusual auroral display.

German astronomer Peter Apian, after seeing five comets in quick succession (the last being the Great Comet of 1540), noted that a comet's tail always points away from the sun.

Queen Elizabeth I, never noted for superstition, commented merely, "Iacta est alea" (the die is thrown), after seeing the Great Comet of 1557.

> Comets, importing change of times and states,
> Brandish your crystal tresses to the sky
> And with them scourge the bad revolting stars
> That have consented unto Henry's death.

Shakespeare, *Henry IV, Part 1*, 1597 A.D.

> When beggars die there are no comets seen;
> The heavens themselves blaze forth the deaths of princes.

Shakespeare, *Julius Caesar*, 1599 A.D.

Johann Cysat, a monk and lens maker, was the first person to scientifically observe a comet through a telescope when he turned his glass toward the Great Comet of 1607. Johannes Kepler telescopically observed the Great Comet of 1618, and proclaimed that comets were prophetic and contained impurities of the atmosphere.

Satan stood
Unterrified and like a comet burn'd
That fires the length of Ophiuchus huge
In Artick sky, and from its horrid hair
Shakes pestilence and war.

John Milton, writing in *Paradise Lost,* after seeing the Great Comet of 1618.

A blazing star
Threatens the world with famine, plague, and war;
To princes, death; to kingdoms, many crosses;
To all estates, inevitable losses;
To herdsmen, rot; to ploughmen, hapless seasons;
To sailors, storms; to cities, civil treasons.

Guillaume du Bartas, *Du Bartas His Divine Weekes and Works,* 1621 A.D.

Daniel Defoe, in his *Journal of the Plague Years,* commented on the appearance of comets in 1664 and 1665:

. . . a blazing star or comet appeared for several months before the plague and then did the year after another, a little before the fire. . .

Astrologer John Gladbury classified various kinds of comets and their effects in 1665:

Dart-like: poor crops, death of kings and nobles.
Red: famine, rising prices, wars.
Rod-like: drought and scarcity.
Hairy-tailed: death of kings, revolution.
Great-tailed: death of notables, war.

Edmund Halley and Isaac Newton observed the Great Comet of 1682. In 1695 Halley calculated its orbit and predicted its return in December of 1758. On December 25th of that year astronomer Johann Palitzch recovered the comet, as Halley predicted.

Charles Messier was the first Frenchman to observe Halley's comet's return. During the course of his career he discovered twenty-one comets, and was referred to affectionately by King Louis XV in 1758 as "my little comet ferret."

Anders Lexell discovered the first short-period comet in 1770. The unusual comet grew in size to five times the diameter of the moon. It is the closest known approach to earth by a comet: distance, 0.015 AU (2.24 million kilometers; 1.4 million miles).

The first woman astronomer of note, Caroline Herschel, discovered eight comets between 1772 and 1848. (Other women astronomers have been immortalized by their comet discoveries, among them: Ludmila Pajdusakova; Liisi Oterma and Eleanor Helin.)

1776: Napoleon Bonaparte is born, and the United States of America is born. The Great Comet of 1776 is held responsible for both these events.

In 1811 two Great Comets appeared, the first discovered by French astronomer Honore Flaugergues. This brilliant comet was visible for seventeen months, and had a tail estimated at 208 million kilometers (130 million miles). Because the French were proud of their countryman, and because 1811 was a particularly good wine year, it is said that this was the year of the great "comet wine."

Jean Pons, former porter, doorkeeper, and janitor of the Marseilles Observatory, holds the record for the greatest number of comets ever discovered: thirty-seven comets between 1801 and 1827.

Great Comet 1843 I, one of the best ever seen, had a tail 320 million kilometers (200 million miles) long and was visible in daylight.

James Howard, third Earl of Malmsbury, upon seeing Donati's comet in 1858 wrote,

The largest comet I ever saw became visible with a very broad tail spread perpendicularly over the sky, the weather being very hot. Everyone now believes in war.

John Tebbutt, an Australian sheep farmer and amateur astronomer, discovered comet 1861 II (some sources mistakenly refer to this comet as 1881 II). The Earth passed through its tail on June 30th, at a distance of 2/3 the length of the comet's tail from its coma. The sky assumed a yellowish, auroral glow; candles had to be lit by 7:00 pm, and the comet was clearly visible by 7:45 pm.

Giovanni Donati observed the first spectrum of a comet and was appointed director of the Florence Observatory in 1864.

In 1881 H. H. Warner made a standing offer of two hundred dollars for discoveries of comets by people in the U. S. A. and Canada. E. E. Barnard discovered nineteen comets and constructed a house, referred to as "the house that comets built."

Sir David Gill photographed the Great September Comet of 1882, the first successful photograph of a comet ever. The first photometric photograph of a comet, by J. Jansen, and the first photographic spectrum of a comet, by Sir William Higgins, were made of the same comet. E. E. Barnard discovered the first comet by photography in 1892.

In 1901 Russian scientist Pyotr Lebedev discovered that light exerts pressure, and deduced why comet's tails always point away from the sun.

In 1909 Samuel Clemens, the author Mark Twain, remarked to a friend,

> I came in with Halley's comet in 1835. It is coming again next year, and I expect to go out with it.

With the expected arrival of Halley's comet in 1910, it was predicted that the earth would pass through the comet's tail. People spent millions of dollars on "comet pills" to ward off the bad effects of the comet. Many people sold off their property and converted their cash to gold, so their wealth wouldn't be consumed by the "cometary fires." Houses were sealed off against noxious fumes of the comet. Other more gullible and unfortunate people went insane or committed suicide.

The Great Comet Kohoutek was discovered in summer of 1973. A religious cult calling themselves The Children of God announced that the comet heralded the end of the world, expected to occur on January 31, 1974.

Halley's comet is returning in 1985-86. What silly and tragic things will people believe this time? Remember, comets are not evil stars; they cannot have any physical effects on people or events. Comets are merely rare and beautiful celestial events, to be experienced and enjoyed!

Unusual Comets

Divided Comets

Many comets have split or become fragmented on their close approach to the sun. This startling event was witnessed by Aristotle as he observed the comet of 371 B.C. The comet of 1618 was observed to fragment by Johannes Kepler. Biela's comet, first seen in 1772, fragmented upon its return in 1845. This is the comet first associated with a meteor shower, the Andromedids.

The most bizarre case of a fragmenting comet occurred in 1882. E. E. Barnard, the noted comet hunter, fell asleep at the observatory while tracking Comet 1882 II. As he slept, he dreamed that he saw the sky "filled with comets." Early that morning the comet split into ten to fifteen pieces!

The most recent comet to split was comet West, in March, 1976.

Colored Comets

The Chinese recorded forty-nine colored comets: twenty-three white; twenty bluish; four red to reddish-yellow; two green.

Seneca reported the comet of 146 B.C. to be "fiery red."

The comet of 1217 A.D. was said to be blue in color, and Pingre reported that the comet of 1476 was "pale blue, bordering on black."

How Comets Are Named

Upon discovery, a comet is named for the person who discovered it, the year it was discovered, and is given a lower-case letter showing its place in the order of comets discovered that year: thus, Comet Kohoutek 1973f was discovered by Lubos Kohoutek in 1973 and was the sixth comet reported that year. Comets discovered by more than one person on the same night are named for all of them (or for the first three, if there are more than that) in the order their reports reach the International Astronomical Union.

Once a comet's orbit is determined it is given a new number, the year and a number (a Roman numeral) showing its place in the order in which comets passed perihelion that year. Thus, Kohoutek's comet is now called Comet 1973 XII, the twelfth comet to pass perihelion in 1973. Sometimes these comets are referred to with their discoverer's names and sometimes not. Frequently historical comets which were particularly notable are called Great Comets. Periodic comets are given the capital letter P, and their discoverer's names, with no year: thus, P/Pons-Winnecke.

There are some notable exceptions. Comets P/Halley, P/Encke, P/Lexell and P/Crommelin are named after the mathematicians who made basic discoveries about their periodicity. Comets called Tsuchinshan are named for the Purple Mountain Observatory in Nanjing, China, where they were discovered, as the Chinese have preferred not to single out individuals.

Comets are sometimes observed by orbital satellites before they are seen by human observers, and the satellites are given due credit: thus, Comet IRAS (Infra-Red Astronomical Satellite)-Iraki-Alcock 1983d.

Eclipse Comets

Several comets have been observed during total eclipses of the sun. Posidonius (135–51 B.C.), writing about an eclipse, recorded that "A comet became visible which had been hidden in its vicinity."

The most memorable eclipse comet was the Tewfik comet of May, 1882, during an eclipse observed from Egypt. Others were observed on July 19, 418 A.D., and November 1, 1948 A.D.

Sungrazers

There is a family of comets that literally travel through the upper atmosphere of the sun. There are at least eleven known periodic sungrazers, ten of which travel similar orbits. The comets in this group of ten are known as Kreutz comets, named after the Dutch astronomer who studied them. Sungrazers are candidates to become divided comets.

Suncrashers

Some sungrazers get too close. The first was discovered by U. S. Defense Dept. satellite P78-1, operated by the Naval Research Laboratory in Washington, D.C. At 21 hours 56 minutes Universal Time, August 30th, 1979, it photographed a comet which barrelled into the sun at 560 km/sec (12,600,000 mph). The comet was independently observed before it crashed into the sun by Howard-Kooman-Michaels, and was designated Comet 1979 XI. The energy output of the comet-Sun collision was measured to be 10^{30} ergs, about a thousand times the total amount of energy used in the United States during a year. Other suncrashers were observed by satellite P78-1 on January 27th and July 20th, 1981.

Earth Grazers

The closest comet to approach Earth was Lexell's comet of 1770, at a distance of 0.015 AU (2.224 million kilometers; 1.39 million miles). Other close encounters were comet IRAS-Iraki-Alcock 1983d, at 4.64 million kilometers (2.9 million miles), and comet Sugano-Saigusha-Fujikawa 1983e, which came within 12.8 million kilometers (8 million miles).

Earthcrashers

It is suspected that the explosion over Tunguska, Siberia in 1908 was caused by either a comet or a meteor—exactly which may never be known. Thunderclaps were heard 960 kilometers (600 miles) away, and the flash was seen in London, England. People and horses were knocked down at a distance of over 240 kilometers (150 miles) from the explosion, and a flame was seen to rise nineteen kilometers (twelve miles) into the sky. An area of Siberian forest ninety-six kilometers (sixty miles) in diameter was flattened.

What Is A Comet, Anyway?

In 1950 astronomer Fred Whipple described a comet as "... a dirty snowball, a ball of frozen ices mixed with dirt, composed of the elements carbon and sulfur, and silicates." But, what is the composition of the frozen ices and snows? Water? Methane (natural gas)? Ammonia? Carbon monoxide and carbon dioxide? Methyl cyanide? Hydrogen cyanide? One compound scientists are having a hard time detecting in comets is ordinary water, and some scientists are questioning whether one of the ices in a comet even needs to be water (See box on following page).

One of the biggest problems in the study of comets is that no one has ever seen the nucleus of a comet up close (in 1983 two comets came within 6.4 million kilometers, or 4 million miles, of Earth, a close approach as comets go). As a result, scientists so far have had to use indirect evidence to determine the makeup of a comet.

The question scientists are trying to answer is: from the material that is detected in the comet's coma and tail, can the original recipe of the snows and ices making up the nucleus be deduced? And what does this tell us, if anything, about the early history of the solar nebula, the Sun, and the planets?

The Nucleus

A comet has a single solid core, or nucleus. The size of the nucleus is estimated to be up to several kilometers in diameter. The nucleus of Halley's comet is estimated to be one to six kilometers (5/8 to three miles) in diameter. However, some astronomers estimate that comets are as much as ten times these dimensions.

The nucleus is composed of ices and dust in a fragile, but stable, crystalline structure, literally a "dirty snowball." In 1975 Fred Whipple suggested that a comet's nucleus is probably a "complex, lacy, fragile structure of ice whiskers and snowflakes." This kind of crystalline structure is called a clathrate.

A comet's nucleus rotates. The rate of rotation is affected by what appears to be explosive evaporation of pockets of ices in the nucleus. The gases escape with a rocket-like action which changes both the rate of rotation of the comet and its orbit.

Observed Chemical Composition of Comets

The Coma

Organic: C, C_2, C_3, CH, CN, CO, CS, HCN, CH_3CN, CH_2O (IRAS 1983d), C_2H_2, $^{12}C^{13}$, S (IRAS 1983d)

Inorganic: H, NH_2, [O], O, OH, H_2O(?)

Metals: Na, Ca, Cr, Co, Mn, Fe, Ni, Cu, V, Si

The Tail

Ions: CO^+, $CO_2{}^+$, CH^+, CN^+, $N_2{}^+$, OH^+, H_2O^+, H_2S^+(IRAS 1983d)

Dust: silicates

These "explosions" also change the visual appearance and overall brightness of the comet.

What is the nucleus really like? Is it layered? or is it pockets or wads of ice and dust? Is it really like a dirty snowball? Or are comets solid asteroidal cores that are coated with ice?

The Coma

The coma is a comet's temporary atmosphere, a swirling mist of dust and sublimated gases that expand away from the comet in the vacuum of space. Because the nucleus of the comet is so small there is no gravity to keep the coma near the nucleus, so eventually the gases and dust will be lost to the void of interplanetary space where they are acted upon by solar radiation and the solar wind.

Solar radiation is energy emitted by the sun in the form of X-rays, ultraviolet radiation (the radiation responsible for your summer sunburn), visible light, and infrared radiation (heat). This energy is emitted in packages which scientists call photons. The solar wind is a steady stream of electrons, protons and other nuclear particles ejected from the sun. Scientists are still searching for the answer to the question, how does solar radiation and the solar wind change the elements that make up the nucleus, to produce the chemistry which we see in the coma?

Comets do not generally form a noticeable coma and tail until they are well inside the orbit of the planet Mars. However, short-period comet P/Schwassman-Wachmann 1, in a nearly circular orbit near the orbit of Jupiter, occasionally fluctuates in brightness, perhaps indicating an active coma. In 1973, when comet Kohoutek was discovered beyond Jupiter, its coma was already active.

In February, 1984, French astronomers lead by Jean Lecacheux used an electronic camera attached to the Hawaiian-Canadian-French telescope atop Mauna Kea in Hawaii to record the brightness of Halley's comet. They were hoping to "see" its bare nucleus while it was 990 million kilometers from the sun (620 million miles, just inside the orbit of Saturn), a distance at which the comet should be inactive. What they actually found was that the comet's brightness was changing. Is this change in brightness evidence of a rotating nucleus, or is it evidence that the volatile ices are already evaporating, even at that huge distance from the sun?

The coma can become enormous. The visible surface of the sun is about 1.4 mil-

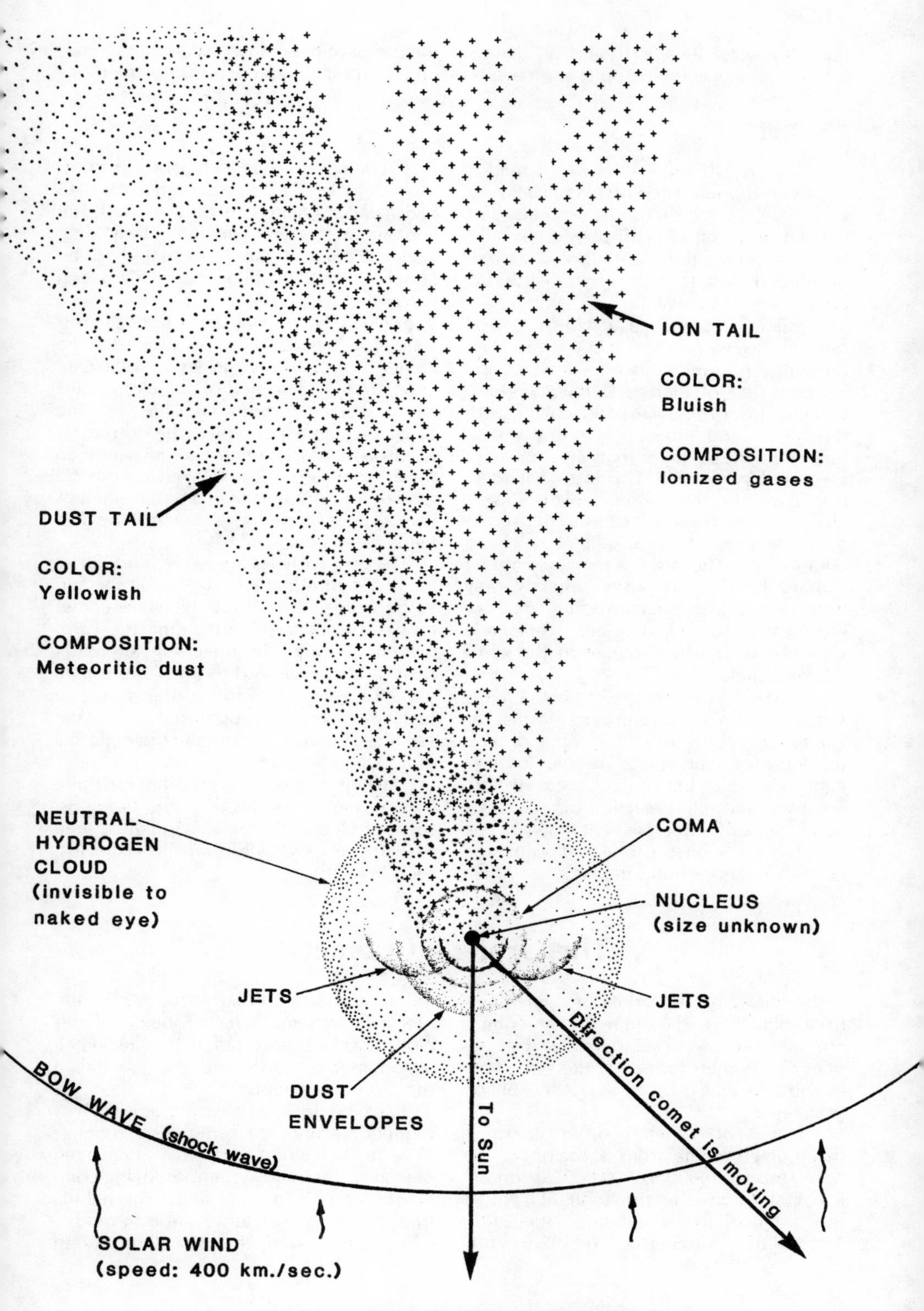

Figure 1. The anatomy of a comet.

lion kilometers (864,000 miles) in diameter, but comets at their closest approach to the sun have been observed to have comas up to ten billion kilometers in diameter!

The Tail

The most striking feature of a comet is its tail. Although some comets display no visible tail, most have one or two tails. The Great Comet of 1744 displayed six tails; Halley's comet does not display a well-developed dust tail. In other words, all comets are not the same.

The Ion Tail (Type I). The bluish-colored ion tail always points directly away from the sun. It is formed by the solar wind, and is composed of ionized cometary gases. Particles from the solar wind collide with the atoms and molecules in the coma, hitting them so hard that electrons are separated from them. This loss of an electron leaves the atom or molecule positively charged, and in some cases the molecules are broken apart to form simpler gases and compounds. The gases in the coma are hit so hard that they are pushed directly away from the sun, and because they are now electrically charged, they are swept away with the solar wind, caught in the sun's magnetic field.

These ionized gases are in a high-energy state. When they return to normal by capturing an electron, they emit or give up some of that energy in the form of light. This emission of light is called fluorescence, the same reaction that causes a neon tube or fluorescent light to glow. The gas that gives off the bluish glow in the comet's tail is carbon monoxide.

A comet's ion tail can grow to enormous length. The Great Comet of 1843 had a tail estimated to be over 320 million kilometers (200 million miles) long. Halley's comet's tail is expected to be about 112 million kilometers (70 million miles) long.

The Dust Tail (Types II and III). This yellowish-colored tail curves away from the comet's ion tail. It is caused by solar radiation, and is composed of particles of dust called silicates. Photons collide with the dust in the coma, and push the dust away. Because the dust particles are heavier than the ionized gases they curve away from the gas tail, but the dust tail can be as long as the gas tail. The yellowish color of the dust tail is reflected sunlight.

The dust particles are left behind, forming a stream of particles that orbit the sun in the same path as the comet. If the comet crosses the orbit of Earth, then the Earth will pass through this meteoroid stream at least once a year. At this time these small particles enter the Earth's atmosphere at high speed and burn up from friction. We see these grains flash through the night sky as a meteor shower.

Halley's comet is responsible for three annual meteor showers: the Orionids, seen on Oct. 20th; the Eta Aquarids, seen May 2nd through 4th; and the Halleyids, seen May 8th.

The Orbits Of Comets

By analyzing the orbits of comets, astronomers have determined that comets are members of the solar system. This implies that comets formed at the same time as the Sun and planets, about 4.6 billion years ago.

Analysis of cometary orbits shows us that comets come from a region of the solar system 20,000 to 100,000 Astronomical Units from the sun (one AU equals the distance from the Sun to the Earth, 149,597,870 kilometers, or 92,956,000 miles), a vast cloud that surrounds the entire solar system. The existence of this cloud was first suggested in 1950 by Dutch astronomer Jan Oort, and it is now called the Oort comet cloud.

Because they originate from such a tremendous distance from the sun, comets must be the oldest and most unchanged objects in the solar system. Studying comets should give us clues about the nebula from which the sun and planets formed.

It is estimated that the Oort cloud

contains somewhere between 100 billion and 10 trillion chunks of frozen ices, or cometary nuclei. Several million of these chunks have already fallen toward the sun. The major cratering that we see on Mercury, the Earth, Earth's moon, Mars, Mars's satellites, and the satellites of Jupiter and Saturn, were probably caused by cometary bombardment during the first half-billion years of the solar system's formation.

Comets are arbitrarily divided into two groups: long period comets, comets with orbital periods longer than 200 years; and short-period comets, comets with periods less than 200 years. One of the longest period comets known is Comet Delevan 1914 V, with an orbital period of 24 million years. The shortest period comet known is P/Encke, with a period of only 3.3 years.

Most comets orbit the sun in the same direction as the planets. If you were to look at the solar system from a position above the Earth's north pole, you would see the planets and comets move in a counter-clockwise direction. This kind of orbit is called prograde. However, Halley's comet orbits the Sun in the opposite direction from the planets, an orbit called retrograde. There are only three other retrograde, short-period comets: Tempel-Tuttle, Pons-Gambert, and Swift-Tuttle. Along with Halley's, they may make up their own cometary family.

The orbit that a comet follows around the sun can be described by three geometrical shapes classed as conic sections. They are elliptic, parabolic, and hyperbolic (See Figure 2).

Elliptical orbits range from the nearly circular, like Earth's, to cigar-shaped, for comets with periods around 200 years. All comets with elliptical orbits have been gravitationally captured (had their orbits radically changed and shortened) by the gas giant planets Jupiter, Saturn, Uranus, and Neptune. In 1886 comet Brooks was observed as it was being captured by Jupiter.

Parabolic and hyperbolic orbits are open-ended, so comets traveling these orbits are not periodic. Comets with parabolic orbits start out from some distant point in the Oort cloud, fall around the sun, and return to some other point in the Oort cloud, but do not return to the sun. Comets with hyperbolic orbits appear to have been deflected from an elliptical orbit by one of the gas giant planets. Comets ejected on hyperbolic paths leave the solar system and the Oort cloud, to become interstellar space travellers. No comet has ever been seen entering the solar system on a hyperbolic path.

In 1979 astronomer Brian Marsden published in his *Comet Catalog* the following statistics about the orbits of 658 individual comets:

1. 275 comets have elliptical orbits.
2. Of the 275, 113 are short period, 162 are long period.
3. 285 comets have parabolic orbits.
4. 98 comets have "slightly hyperbolic" orbits.
5. No comets observed have "strongly hyperbolic" orbits.

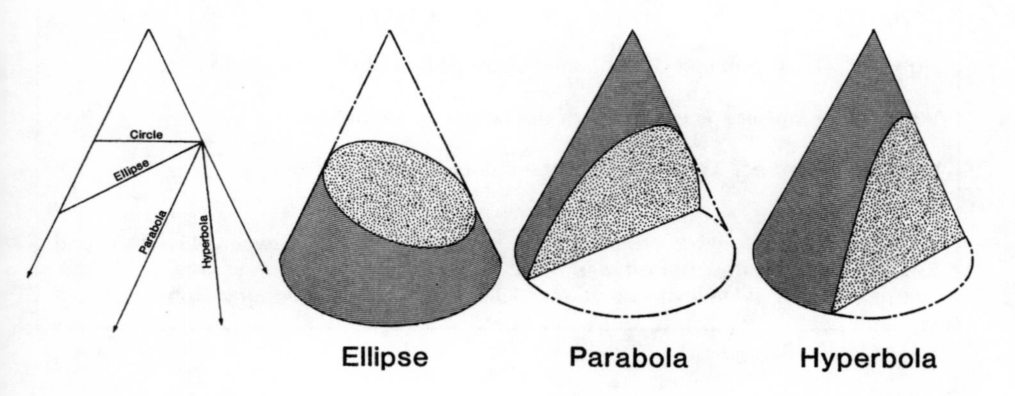

Figure 2. How to cut a cone to make conic sections.

Halley's Comet's Orbital Elements

(from the *Comet Halley Handbook* by Donald Yeomans)

Halley's comet's orbit is computed from a total of 885 observations. The first observation was made on Sept. 28, 1607, by Johannes Kepler; the last was made on May 24, 1911.

T = 1986 Feb 9.6613 Ephemeris Time. This is the moment when the comet will be at its closest approach to the sun, called perihelion.

q = 0.587096 AU. This is the comet's perihelion distance, measured in Astronomical Units.

e = 0.967267. This is the comet's orbit's eccentricity. This number tells you how far the comet's orbit is from being a circle (e = 0) or a parabola (e = 1).

☊ = 58.1531°. The Ascending Node. This is the point of intersection between the plane of the orbit of the comet and the plane of the orbit of the Earth, on the comet's south-to-north crossing.

☋ = 238.1531°. The Descending Node. This is the point of intersection between the comet's orbit and the Earth's orbit on its north-to-south crossing.

ω = 111.8534°. Argument of Perihelion. This is the angle measured from the position of the ascending node to the perihelion position of the comet, measured in the plane of the orbit of the comet, in the direction of the comet's orbit.

i = 162.2738°. Inclination. The angle the comet's orbit makes with (or is inclined to) the orbit of the Earth.

Q = 35 AU. Aphelion distance. The comet's farthest distance from the sun.

Z(Q) = 9.9 AU. The distance the comet travels below the orbital plane of the Earth.

z(q) = .17 AU. The height of the comet above the orbital plane of the Earth.

V(q) = 54.55 km/sec. The velocity of the comet at perihelion.

V(Q) = 0.91 km/sec. The velocity of the comet at aphelion.

Halley's comet's longest orbital period was 79.25 years, between 451 A.D. and 530 A.D. The shortest period was 74.42 years, between 1835 A.D. and 1910 A.D. The next return of Halley's comet after 1986 is predicted to be June, 2061.

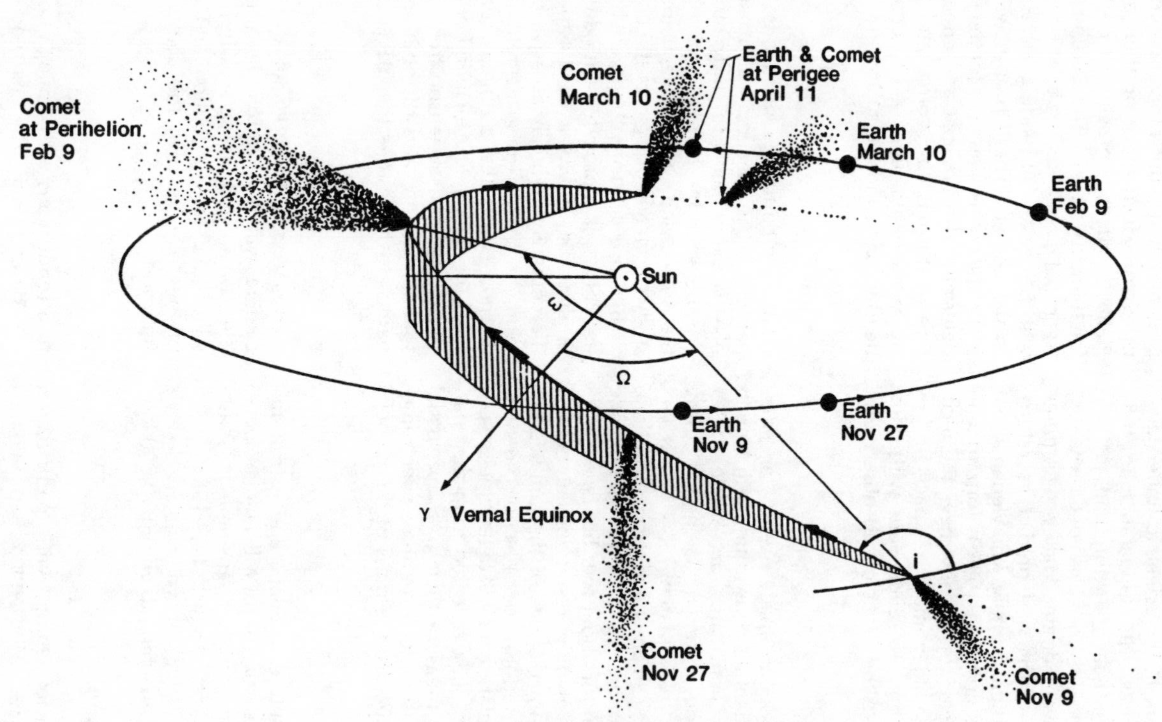

Figure 3. Orbital elements of Halley's comet.

Space Probes To Halley's Comet

The appearance of Halley's Comet in 1986 marks its second occurrence as a turning point in the scientific understanding of comets. Its appearance in 1682 came when Isaac Newton had just figured out the Universal Laws of Gravitation, which allowed Edmund Halley to compute the comet's orbit and predict its return, the first certain scientific knowledge about comets. Its arrival in 1986 is fortuitous in the same way, for we now have the technology to build unmanned space probes and send them to rendezvous with comets, to make the first closeup observation of a comet.

The United States's space technology is unmatched. But other countries want to explore deep space on their own, and Halley's comet is proving to be the spur for them to do so. The European Space Agency, Japan, and the Soviet Union along with France are all sending space missions to the comet. These probes will photograph the comet's nucleus for the first time, determine the physical and chemical nature of the nucleus, coma, and tail, and study the interaction of the comet with nearby space and the solar wind.

Giotto

The European Space Agency, an international scientific consortium, will launch spacecraft Giotto atop an Ariane rocket from Korou, French Guinea, during July, 1985. On March 13, 1986, it will reach Halley's comet and, from a distance of 500 kilometers (312 miles) send back information for about four hours.

Giotto's medium-resolution camera will take close-up pictures of the comet's nucleus and perform spectroscopic analysis. Its photopolarimeter will determine the polarization of light reflected from dust particles, and measure their size and spatial density as well. Its three mass spectrometers will measure the mass of the comet's neutral gas particles, plasma ions, and dust particles. The spacecraft also has several plasma experiments to study the interaction between the comet's ionosphere and the solar wind, and a magnetometer to study the magnetic field of the comet.

Because Giotto will approach close to the comet, it is provided with dust shields to allow it to survive as long as possible in the comet's potentially abrasive dust environment. In fact, the spacecraft's designers do not expect the probe to survive beyond the encounter. Included on the dust shields are acoustic sensors to measure the speed, position, and distribution of dust particles which strike the spacecraft.

Planet-A

The Institute of Space and Astronomical Sciences of Tokyo will launch spacecraft Planet-A from Kagoshima in southern Japan on August 14, 1985. On March 8, 1986, it will reach Halley's comet, approaching to within 10,000 kilometers (6,000 miles). Planet-A's ultraviolet camera will photograph the structure of the coma, and its solar wind analyzer will study the distribution and direction of the solar wind in the comet's neighborhood.

Vega 1 & 2

The Soviet Union, with participation from France, will launch two identical spacecraft, Vega 1 & 2, two weeks apart in December, 1985. After dropping two small probes at Venus, the main probes will use Venus's gravity as a sling to send them on to a rendezvous with Halley's comet during March, 1986. The first spacecraft will be aimed to pass 10,000 kilometers from the comet's coma. Two weeks later the second spacecraft will arrive, and may be sent closer. Each spacecraft is expected to send data for about three hours.

Each Vega carries two cameras, a wide-angle camera to photograph the coma, and a narrow-angle camera to photograph details of the nucleus as small as 200 meters across from a distance of 10,000 kilometers. The spacecraft have spectrographs to determine the composition and rate of outflowing gas, and the polarization of light from dust particles; and camera/spectrographs to take infrared pictures of the nucleus, determine its temperature, and determine the nature, content, and temperature of the comet's dust and gas.

Like Giotto, the Vegas carry plasma experiments and magnetometers. They also have dust shields with impact sensors, and at least the closer spacecraft of the pair is not expected to survive the encounter.

Only these groups will be sending unmanned space probes to Halley's. The United States is settling for second-best, taking a back seat to the space efforts of these other national and international space programs. Unlike the detailed, live pictures we have grown accustomed to seeing from our own space efforts, unmanned as well as manned, the United States will be relegated to the role of information clearinghouse and armchair audience.

Although planning for sophisticated spacecraft and science missions was begun by NASA in the mid-seventies, the United States government decided in 1981 to cut off all funding for a Comet Halley mission. This is most unfortunate, as the United States has the only space program with the technology and proven experience to build a long-lived spacecraft able to keep up with the incredible speed of a comet and to make high-resolution pictures of a comet. Also, the U. S. is the only country with a deep-space tracking network capable of virtually round-the-clock contact with deep-space probes.

The United States does have its little finger in the cometary pie. The International Halley Watch, an information clearinghouse and coordinating body for information from near-Earth orbit and deep-space probes, is being operated by the U. S.

The United States is also operating its own alternate, shirttail comet mission. The International Sun-Earth Explorer (ISEE)-3 satellite has been sent away from its prime mission of observing the sun, onto a trajectory which will take it near comet P/Giacobini-Zinner in September, 1985, to study the interaction between that comet and the solar wind. It will also make long-distance observations of the solar wind near Halley's comet in 1985 and '86. However, this spacecraft is not designed to make any basic, comet-specific observations, and it has no camera.

Joseph Veverka, a comet scientist from Cornell University who has likened the US's nonsupport of a primary Halley's comet mission to a national disaster, has stated,

> If the United States doesn't make the effort to explore the most famous of all comets, our children may not forgive us.

Calibrating Your Fist

The moon is 1/2° wide in the sky. To make measuring distances easier than trying to count moons between star and comet, you can use your hand or fist held at arm's length. The pointer stars in the Big Dipper are 5° apart. With your arm fully extended see how many fingers fit between the pointer stars. That distance, 5° of sky measured by your fist, is easy to remember and a handy measuring tool.

A Halley's Comet Observing Calendar

October '85: Telescope And Large Binocular Viewing

Moon Phase:		
	Last quarter	6th
	New	13th
	First quarter	20th
	Full	28th

Velocity of the comet increases from 25.27 km/sec on the 1st to 29.42 km/sec on the 31st.

At the beginning of the month be set up and ready for viewing by midnight, but by the end of the month you should begin viewing by 10:00 pm.

The comet begins its retrograde sweep across the sky, entering the constellation Taurus on the 24th. Comet-Earth distance shrinks from 2.04 to 1.08 AU, and Comet-Sun distance shrinks from 2.35 to 1.94 AU.

At the beginning of the month look near Orion's club; on the 20th look 13 1/2° above the star Betelgeuse, the bright red star of Orion's shoulder. On the 28th the comet will pass less than 1° below the Crab Nebula, M-1. On the 31st the comet will pass 5° below the star Elnath (the second brightest star in Taurus) and 16° above the star Bellatrix, the star of Orion's other shoulder.

Orionid Meteor Shower. The Orionid meteor shower peaks on the morning of the 20th. These meteors are dust from previous passages of Halley's comet. The average rate is twenty-five meteors per hour. If you trace the meteors back to their point of origin they will appear to radiate from Orion's club.

November '85: Binocular Viewing

Moon Phase:		
	Last quarter	5th
	New	12th
	First quarter	19th
	Full	27th

Velocity of the comet increases from 29.25 km/sec on the first to 33.68 km/sec on the 30th.

Best viewing will be the first two weeks of the month at midnight.

The comet continues its retrograde sweep across the sky, entering the constellation Aries on the 18th and Pisces on the 27th. Comet-Earth distance shrinks from 1.08 AU to 0.63 AU; Comet-Sun distance shrinks from 1.94 AU to 1.5 AU. On the 27th the comet is at its closest point to the Earth (0.62 AU) on its sunward leg; however, the moon is full and will interfere with observing.

On the 10th the comet will pass 5 1/2° above the brightest star in Taurus, Aldebaran. On the 16th the comet will pass about 2° south of the Pleiades. On the 25th the comet passes 7° south of the second brightest star in Aries, Hamal, and on the 27th it will be 4° below the brightest star, Sheratan.

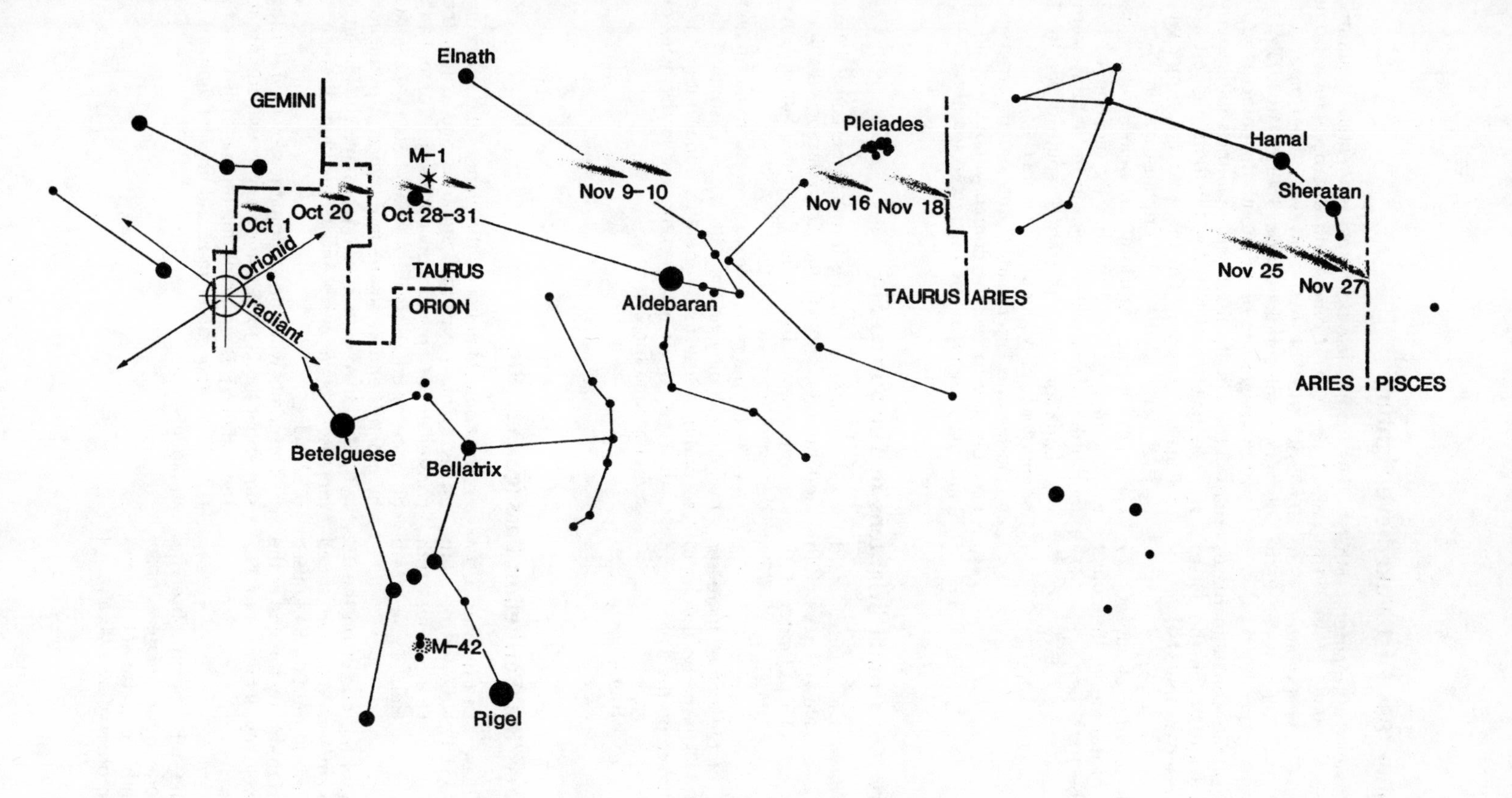

Figure 4. Position of Halley's comet relative to the stars, October and November 1985.

Comet and stars precessed to epoch 1950.

December '85: First Naked-eye Sighting

Moon Phase:

Last quarter	5th
New	11th
First quarter	19th
Full	26th

Velocity of the comet increases from 33.8 km/sec on the 1st to 40.92 km/sec on the 31st.

Be ready to view at sunset. By the end of the month the comet will set by 10:00 pm.

The comet leaves Pisces and enters Aquarius on Dec. 20th. Comet-Earth distance grows from 0.63 AU to 1.14 AU. However, the Comet-Sun distance is still shrinking, from 1.49 AU to 1.03 AU.

At the beginning of the month the comet is 130° east of the sun at sunset, visible in the southeastern sky. On the 6th look 6° below the third brightest star in Pegasus, Algenib. On the 12th the comet will be 1/2° north of the third brightest star in Pisces, Gamma. On the 16th look 11° below the brightest star in Pegasus, Markab. By the end of the month the comet will be seen in the southwestern sky. Look for it on the 30th, 1° south of the third brightest star in Aquarius, Sadachbia.

During the latter part of the month observers may be able to see the comet without the aid of binoculars. Who will be the first to see it?

January '86: Comet Brightens In The Sunset Sky

Moon Phase:

Last quarter	3rd
New	10th
First Quarter	17th
Full	25th

Velocity of the comet increases from 41.43 km/sec on the 1st to 53.06 km/sec on the 31st.

Be ready to observe right after sunset.

The comet is moving through the constellation Aquarius all month. Comet-Earth distance is still increasing, from 1.16 to 1.56 AU. Comet-Sun distance decreases from 1.01 to 0.62 AU.

Look for the comet near the crescent moon and the planet Jupiter on the 13th. Look 1° south of the second brightest star in Aquarius, Sadalsuud, on the 24th. The comet will become lost in the glare of sunset around the 25th.

February '86: Perihelion Passage Feb. 9th

Moon Phase:

Last quarter	1st
New	8th
First quarter	16th
Full	24th

Velocity of the comet on the 1st is 53.06 km/sec. Maximum speed is reached on the 9th, 54 km/sec (122,373 mph). By the end of the month velocity decreases to 49.7 km/sec.

After the 9th, begin observing about one hour before sunrise. By month's end be set up and ready to observe by 4:00 a.m.

The comet, hidden by the glare of the sun, continues to move through Aquarius and enters the constellation Capricorn on the 23rd. Comet-Earth distance again decreases, from 1.56 to 1.29 AU by month's end. Comet-Sun distance at perihelion is 0.59 AU. After perihelion passage the distance increases; by month's end Comet-Sun distance is 0.82 AU.

The comet will emerge from the glow of the rising sun about the 20th. Some observers will attempt to see if Halley's is visible during the day. The comet's tail will be noticeably larger and brighter than it was in January.

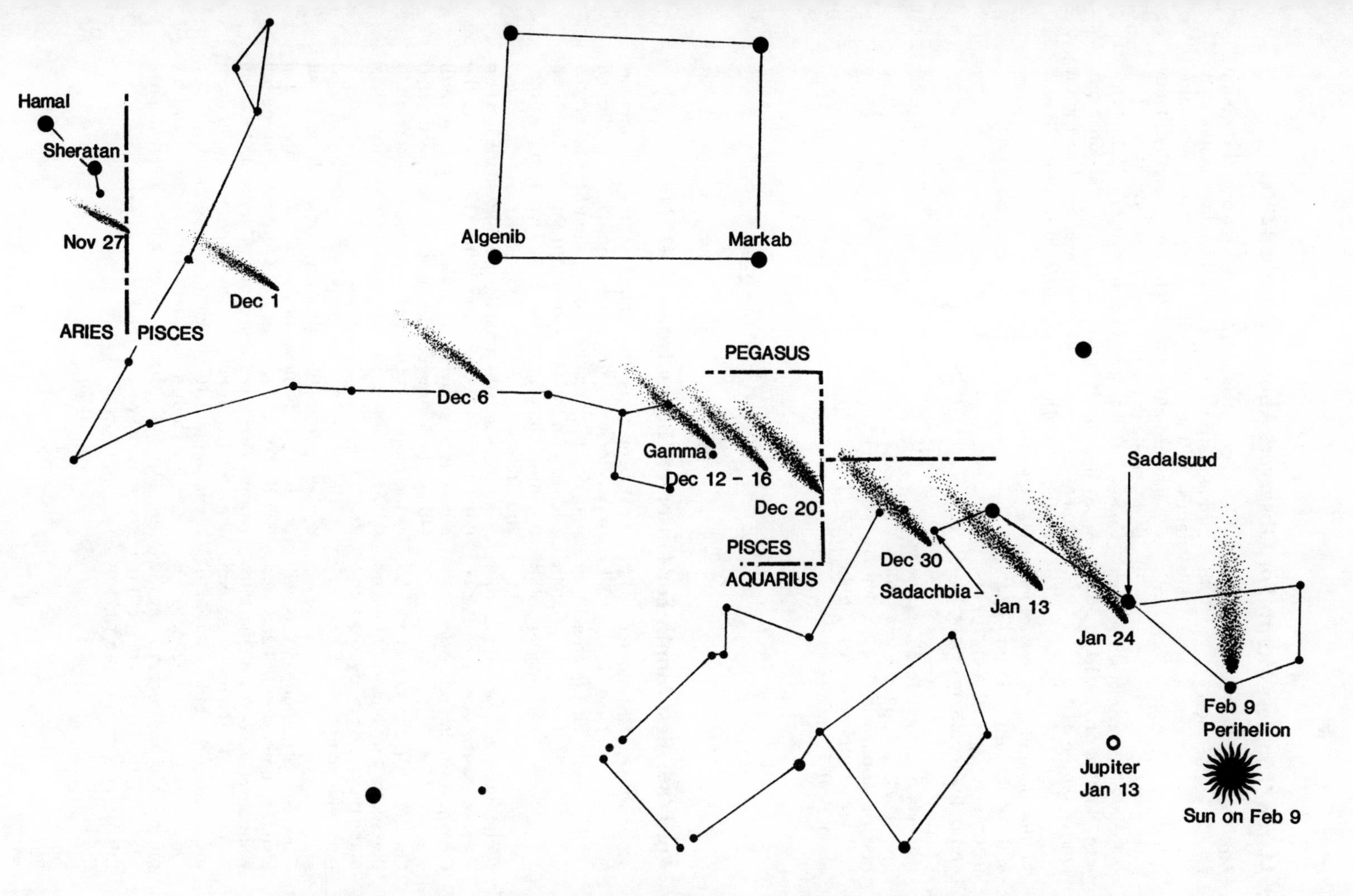

Figure 5. Position of Halley's comet December, 1985, January, 1986, to perihelion February 9th., 1986.

March '86: Best Month For Observers Above Latitude 42° N.

Moon Phase:	Last quarter	3rd
	New	10th
	First quarter	18th
	Full	25th

Velocity of the comet slows from 49.17 km/sec on the 1st to 38.32 km/sec on the 31st.

At the beginning of the month you should be set up and ready to observe by 4:00 a.m.; by the end of the month, midnight.

If you are at latitude 55° N the comet will disappear below your horizon about March 29th, and will not reappear above it until about April 18th.

The comet continues its retrograde motion through Capricorn. It enters the constellation Sagittarius on the 12th, and enters the constellation Corona Australis on the 30th. Comet-Earth distance dramatically decreases to 0.59 AU. on March 31st. Comet-Sun distance grows from 0.72 to 1.17 AU.

As the comet moves away from the sun North American viewers will see it lower and lower in the eastern sky. The tail will grow to between 20° and 30° in length.

April '86: Best Month For Observers Below Latitude 41° N.

Moon Phase:	Last quarter	1st
	New	8th
	First quarter	17th
	*Full	24th
	Last quarter	30th

*Total lunar eclipse, visible from western North America, Pacific Ocean, Australia, S.E. Asia.

On the 11th, the comet will be due south at midnight. At the end of the month, be ready to observe right after sunset.

Observers at latitude 50° N will lose the comet on April 2nd, and will not see it again until the 15th. At latitude 45° N observers will lose it from the 6th to the 12th. Observers at latitude 42° N lose it on the 9th, and regain it on the 11th or 12th. Observers south of this latitude do not lose sight of the comet.

Movement of the comet in the sky appears to accelerate as the comet approaches outbound perigee. It enters the constellation Scorpio on the 2nd, Ara on the 6th, Norma on the 7th, Lupus on the 9th, Centaurus on the 12th, Hydra on the 18th, and Crater on the 25th.

Comet-Earth distance decreases from 0.53 AU on the 1st to 0.42 AU on the 11th. On that date the comet is at perigee, its closest point to the Earth (48 million kilometers; 39 million miles). By the end of the month Comet-Earth distance will increase to 0.80 AU. Comet-Sun distance increases from 1.17 to 1.63 AU.

The comet will be easily seen with the unaided eye away from city lights; the coma will be about one-half the diameter of the full moon, and its tail will be 20° to 40° in length. On the day of the full moon viewers west and south of North America will see a total lunar eclipse. The comet will be 40° from the moon and 100° from the Sun at dawn.

As the Earth moves between the comet and the Sun (April 10-18) the comet will appear to pivot counterclockwise in the sky.

May '86: Binocular viewing

Moon Phase:	New	8th
	First quarter	16th
	Full	23rd
	Last quarter	30th

Velocity of the comet decreases from 32.25 km/sec on the 1st to 28.43 km/sec on the 31st.

Be ready to observe the comet after sunset in the evening sky.

On the 3rd Halley's comet re-enters the constellation Hydra, and by the 12th enters Sextans. Comet-Earth distance increases from 0.08 to 1.80 AU. Comet-Sun distance increases from 1.63 to 2.07 AU.

Halley's comet will fade from view to the naked eye sometime during this month.

Eta Aquarid and Halleyid meteor showers. These showers should be more spectacular this year than normal. Eta Aquarids are visible from May 2nd to 4th; normal rate is twenty meteors per hour. The Halleyids peak on May 8th and observing conditions are perfect with a new moon. Best time to observe meteor showers is after midnight.

Comet Observing Tips And Techniques

A Great Comet, a bright comet with a well-developed coma and tail, is one of nature's most spectacular sights. Halley's comet is a well-known Great Comet, and will be visible to anyone who is drawn to see and appreciate this thrilling sight.

For a good experience you need:

- to dress warmly.
- to select a dark viewing site.
- binoculars with tripod and binocular clamp.
- a flashlight with a red filter.
- a planisphere or easy-to-read star charts.
- the comet observing calendar.

Halley's comet will be best for viewing in the northern hemisphere during the month of March. However, the best single day for the comet will be April 11th, when it is closest to the Earth, 48 million kilometers (39 million miles) away. Around this date the comet's tail will be at its longest, stretching across 20 to 40 degrees of sky. Unfortunately, in April, northern hemisphere viewers will see the comet quite low in the sky, and observers farther north than 40° won't see it at all. (The best observing site in the world on April 11 will be Alice Springs, Northern Territories, Australia, at a latitude of 23° S. The comet will be directly overhead, and the moon will be two days past new.)

When you go out at night to observe, dress warmly. Halley's will be visible January, February, March and April; winter in the northern hemisphere. Pay attention to protection of your feet and face. Overdress; it is always easier to take off a coat or sweater, if you become too warm, than it is to put on more clothes if you didn't bring them. Listen carefully to your local weather forecasts for temperature and cloud cover. Otherwise, you may be a candidate for frostbite, hypothermia, or just disappointment.

Select a good observing site. You need a clear, dark view of the southeast, south, and southwest horizon. Light pollution drastically reduces your ability to see astronomical objects. It is predicted that Halley's comet will not be as bright in 1986 as it was in 1910. If you do not get sufficiently far from the glare of city lights, you may not be able to see Halley's comet at all!

There is no hard-and-fast rule as to how far away from the city you need to be to secure a dark site; your eyes will show you. If you can see the Great Nebula of Orion (M-42), the Milky Way, the Great Galaxy of Andromeda (M-31), or any other deep-

Visibility of Halley's Comet

(Compare these star charts to a planisphere)

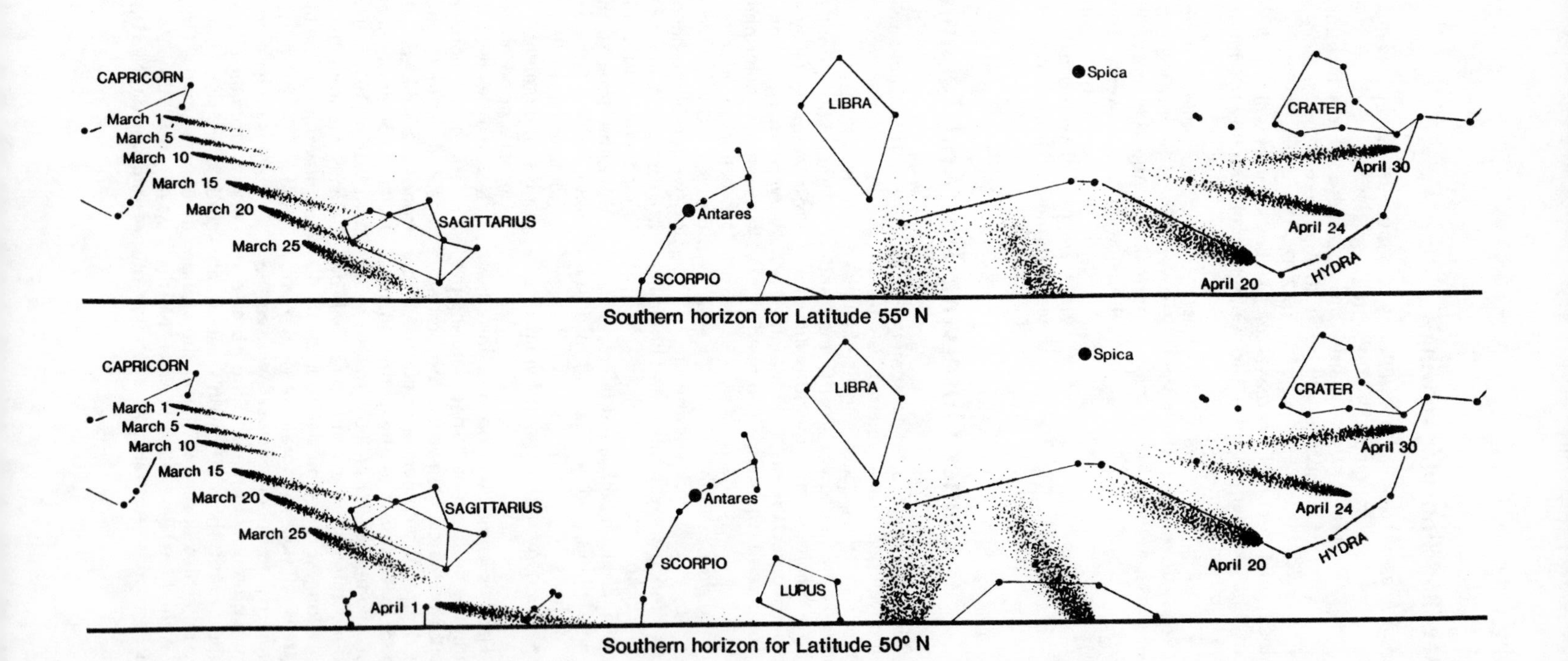

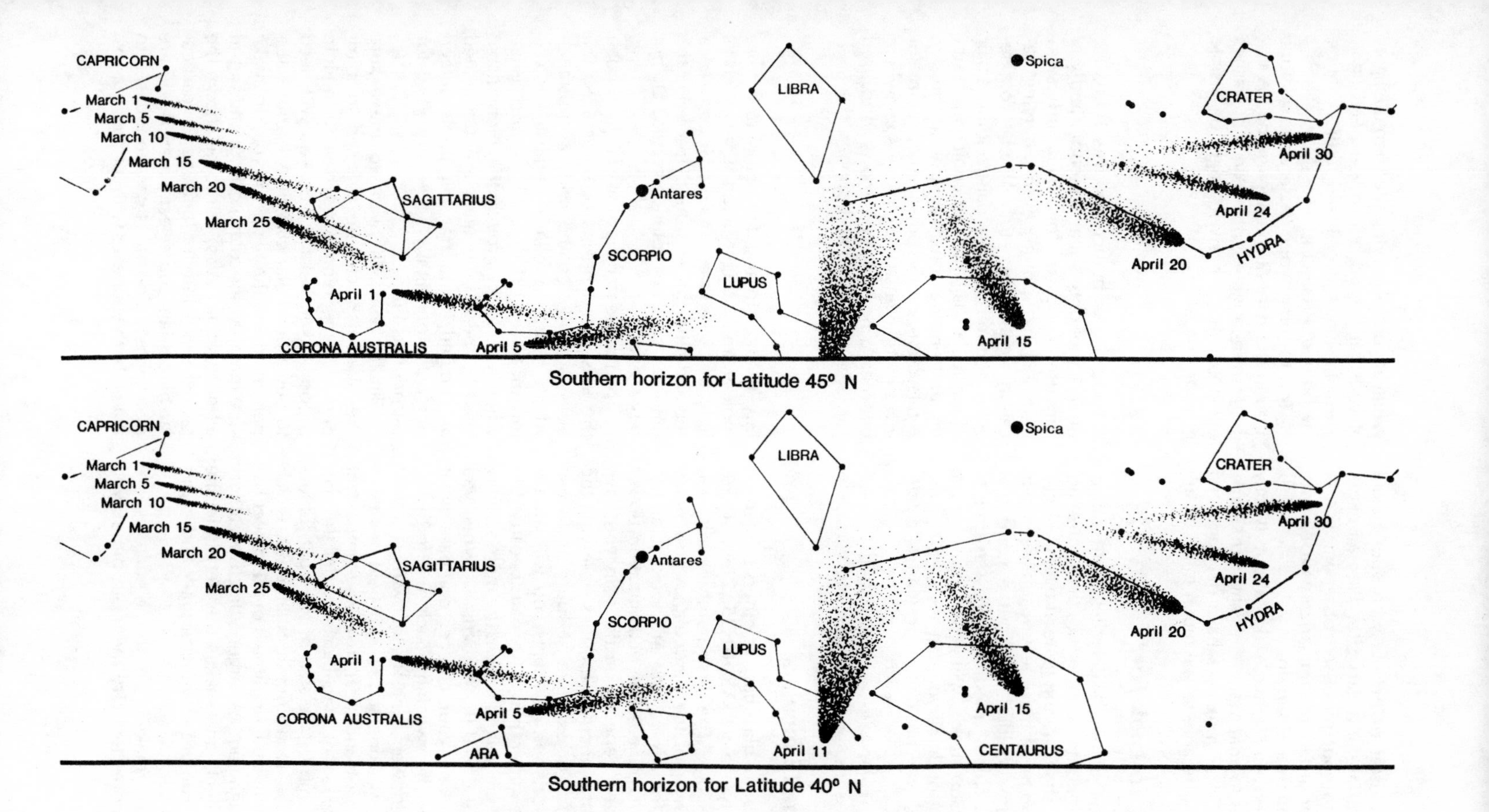

Figure 6. Position of Halley's comet March and April, 1986.

space cluster or nebula with your naked eye, you are at a dark site. Plan an expedition to your proposed observing site in summer or fall, to familiarize yourself with its location and terrain.

When you arrive, take about twenty minutes to adapt your eyesight to the darkness. Once your eyes have dark-adapted you will be able to see much more detail. While at your observing site keep the lights of your car turned off. If you need light, use a flashlight with a red filter covering the bulb and reflector.

Try to plan your observing sessions around the date of the new moon. A visible moon brightens the sky and washes out detail that otherwise might be visible to your eye.

What To Look For

Bright comets like Halley's are active. Look for jets, concentric rings, the nucleus's shadow, and general overall brightness. Some of these features may be seen with the unaided eye, but a good pair of binoculars, mounted on a tripod to hold them steady, will greatly increase the amount of detail you can see.

But, why binoculars, instead of a telescope?

Binoculars are recommended for several reasons, primarily cost and versatility. A good pair of binoculars costs 1/4 to 1/2 the price of only a fair telescope, and you can use binoculars for other purposes—birdwatching, sporting events, hunting, vacation sightseeing, etc. Also, if you develop a serious interest in astronomy, a pair of binoculars is your best choice for learning and finding your way around the sky. A telescope, on the other hand, is a major investment whose cost ranks somewhere between a home stereo and a good used car ($250 to $3,000 and up).

Buying Binoculars

All binoculars have numbers printed on them, typically 7x35, 10x35, 7x50, 10x50, or 20x80. The first number tells you the power of the binoculars: how many times the object you are looking at has been magnified. The magnification power you need depends on the uses you have for your binoculars (besides observing the comet). For example, if you plan to use them for sporting events you will want a 7x pair rather than a 10x pair, because the higher the magnification, the more steadily you must hold the binoculars. When you are making your choice, compare what you see by switching back and forth between powers, and try following a distant person walking across your field of view.

For astronomical use the important number is not magnification, but the second number: the size of the front (objective) lens, measured in millimeters, which determines the brightness of the object being magnified. A larger objective gathers more light, and makes the details in the image you are viewing brighter and easier to see.

Next, make sure the binoculars are color-corrected. Try them out before you buy them. To test for color-correction, look at small, bright light sources such as light bulbs or street lamps. If you see a red-and-blue circle around the light, the binoculars are not color-corrected. Do not buy them.

To improve transmission of light through the lenses and prisms of binoculars, the glass is thinly coated with a substance which reflects a bluish, reddish, or yellowish tinge. (This color is *not* visible when you look through the binoculars.) Usually these binoculars are designated as "coated," and transmit up to fifteen percent more light. These are ideal for astronomical use.

Finally, you will want to make sure the binoculars can be attached to a tripod. Some binoculars have a built-in tripod socket, but for most you will need to buy a binocular clamp. Ask your salesperson to set them up for you, to make sure that the binoculars, clamp, and tripod all fit together. Also, ask to be shown the proper procedure for focusing the particular binoculars you select. Even if you are nearsighted, binoculars can be focused to match both your eyes without your glasses.

Observing With Binoculars

Look for detail in the coma. The central part of the coma will be bright. Surrounding the bright central region will be a diffuse globe, the comet's gaseous atmosphere. Extending into the coma from the bright region will be a series of jets. The jets will curve away from the center of the coma, sweep around it, and disappear into the comet's tail. Jets are not a permanent feature, and may change during several minute's observing. The part of the coma that is closest to the sun will look like a series of concentric, parabolic envelopes around the coma. Much of this detail, easily visible through your binoculars, will not be visible in photographs of the comet. Try drawing what you see.

Look for detail in the structure of the tails. Halley's is well known for its long, thin gas tail, and is noted for having only a weak dust tail. Look for a dark streak down the middle of the gas tail, a streak which may be the shadow of the nucleus itself, cast upon the tail. Careful observations of the tail may reveal a rippling motion as the jets of the nucleus are swept back by the solar wind. A series of photographs taken several minutes apart revealed this rippling motion in comet Kohoutek in 1973–4. Also, look for condensations or the appearance of bright patches forming in the tail. A comet is a very sensitive solar windsock.

Halley's comet is sufficiently active that you may see details change in the coma and tail in observations as close as ten minutes apart. What you see will depend upon the strength of the solar wind and the activity of the comet's nucleus. You may see the bright jets appear and change as volatile material in the nucleus evaporates explosively. You may notice an increase in overall brightness because of increased solar wind and radiation caused by a flare on the sun. It is this kind of change in activity which frightened our ancestors when they observed comets.

So You Still Want To Buy A Telescope

Telescopes, unlike binoculars, are specialized scientific instruments. One advantage of a telescope is that you can change magnification. Another is that telescopes generally have better light gathering capabilities than a pair of binoculars. What kind of telescope you buy depends upon what you want a telescope for, and how much money you have to spend.

Of the four basic categories of telescopes (refractor, Newtonian, Cassegrain, and rich-field reflector), the rich-field telescope is probably the best for observing comets. It sees a large area of sky, and it gathers many times more light than binoculars, as the mirrors of RFTs usually range from four to ten inches in diameter. Price varies from about $250 to $750 and up.

Buying a telescope is much more chancy than buying binoculars. Study the articles and ads in *Sky and Telescope* and *Astronomy* magazines. Beware of extra-cheap telescopes, and ads which claim fantastic magnification ability.

Contact your local amateur astronomy club or planetarium. You may be able to see many different telescopes, and have a chance to look through some of them. The Astronomical League, PO Box 12821, Tucson, AZ, 85732; or the Royal Astronomical Society of Canada, 124 Merton St., Toronto, Ontario, M4S 2Z2 should be able to put you in touch with the amateur club nearest you.

How To Photograph A Comet

A comet is at once one of the easiest, and one of the most difficult, of all astronomical phenomena to photograph. After all, if you can see it you can photograph it, and you can even take its picture with a box Brownie. But, there are better ways.

First, there is one thing to keep in mind. Halley's comet returns only once every seventy-six years, and this time around there may be thirty days when the comet can be photographed easily. Don't waste any of them. Film is cheap on a seventy-six year timescale; when you photograph the comet shoot many frames, bracket your exposures, record your exposure data, and get your film developed quickly. Use the results of one evening's or morning's shoot as the basis for the next, and keep at it.

What You Need

The basic equipment for good-quality comet pictures consists of:

1. Camera with "B" (bulb) setting
2. Tripod
3. Cable release
4. Film
5. Flashlight

To give you better pictures, add:

1. Camera with "B" or "T" (time) setting, any format
2. Locking cable release
3. Lens of f/2.8 or faster
4. Fast black-and-white or color film
5. Flashlight with red filter

The Simplest Comet Pictures

Photograph Halley's comet at dawn or dusk, when the sun is below the horizon. Use film with speeds between ASA/ISO 25 and 400. Your camera should be placed on a steady surface—a tripod is best. Point your camera at the comet, focus, open the lens aperture wide open, and bracket—that is, make a series of exposures. Start with one second, then make exposures increasing by three seconds each time until you have reached thirty seconds. You will need to use the manual override setting of an automatic camera.

You will get a picture of the comet. Depending upon your lens and film, some frames will be too dark, and some too light. Several, however, will be just right, and these are the exposures you can start with at the next evening's or morning's session.

What you will *not* get—and this is very important—is a picture like the one on the cover of this handbook, a picture made with special scientific film and special exposure and guiding techniques, through one of the larger telescopes in the world.

There is no reason to be discouraged. Among your frames you *will* have a good picture of the comet. And there are techniques and equipment, easily available to the photo hobbyist, to give you better pictures, and some techniques to make those pictures uniquely your own.

Size

It seems reasonable to think that, if the comet is too small on the frame, using a telephoto lens will make it bigger. And, yes, that is the case. On a 35mm camera, lenses of 85mm to 135mm focal length (covering an angular diameter from 28° to 18° respectively) will probably fill the frame with the comet. But, changing your lens can cause you another problem.

The Earth turns. This motion causes phenomena familiar to us—the passage of the stars over our heads at night, the rising of the moon, the setting of the sun. The comet will appear to move at the same rate. As you make longer and longer exposures, you will see that the comet, and any stars in your frame, are no longer distinct. As the earth (and your camera) turns under them, they begin to show up as lines, in the case of stars, or a smudge, in the case of the comet.

When you photograph Halley's comet at its brightest (March and April, '86) the comet will be low in the sky. With a "normal" focal length lens (50mm on a 35mm camera) star trailing and comet trailing will not begin to be noticeable until your exposure time approaches thirty seconds. However, with a 135mm telephoto, you may see star trailing in an exposure of only ten seconds. If you consider that it is common for normal lenses to be as fast as f/1.8, and that most medium telephotos are no faster than f/2.8, you see that with a telephoto it may take more than twice the amount of time to get the same exposure on your film. It might be a better idea to shoot with your normal lens and enlarge the picture.

Film Selection

The three types of film that you are most likely to use to photograph the comet are:

1. Black-and-white negative film
2. Color print film
3. Color slide film

Unlike the days when film was slower than ASA 100, and the only color film available was ASA 18, all three kinds of film are now available in speeds as slow as ISO (ASA) 25 and as fast as ISO 1600. As a comet photographer your choice is a compromise between speed (ISO 1600 is six f/stops faster than ISO 25) and graininess (faster film is grainier).

Black-and-white negative film is the most versatile of the three film types. In general it has more latitude (tolerance to under- or over-exposure) and it is by far the easiest to develop and print yourself. This allows you the maximum control over picture contrast, and therefore over the amount of detail you see in the final print. Another plus for black-and-white film is speed of processing. If you develop your own film you can see your results right after your shoot, so you know what to do differently next time.

Color print film is the next most versatile. While it does not generally have any latitude for under-exposure, it has lots for overexposure. Processing speed is its biggest downfall. It remains to be seen whether or not one-hour processors are of sufficient quality for comet photographs. Even if the negatives are okay, one-hour process prints of the comet more than likely will be awful, though they will be useful as an exposure guide. The contrast range of a comet picture virtually demands custom color printing to get black blacks, good contrast, and details in the comet. Most good color print processing takes several days.

For many people, color slide film is the most pleasing. Its biggest drawback in comet photography is its considerable contrast (particularly in prints made from slides), loss of detail in overexposed highlights, and the complete inability to adjust contrast once the film has been exposed. This is the film of choice for many amateur astrophotographers, however, and in many cities you can get same-day or next-day processing for many slide films.

Many readers undoubtedly have videotape equipment. Though there is no data available, try videotaping the comet with your lens wide open, and the camera at its most sensitive setting. Video is the most immediate medium for seeing your results; simply look in the monitor or viewfinder, or play back your tape.

Comet Photography Tips

In general, ISO 200-speed color film and ISO 400-speed black-and-white film are the best compromises for grain, speed, contrast, and detail. Some general tips, which apply to all film, are:

1. Shoot short rolls (twenty to twenty-four exposures). This should cover all your bracketing for a morning or evening shoot, and won't leave you with lots of unexposed frames left over. If you want to shoot more, you can always use another roll.

2. Process promptly, despite any unused frames on the roll. Halley's is a time-critical phenomenon—it won't be back for a long time. Look at your pictures, and use the best exposures as a guide for your next shoot. And, take pictures at every opportunity. Each time out will give you a different picture.

3. When you first load your film, take a picture in full light, even if it is just your flashlight pointed into the lens. Then insist that your processor *not* cut your film. Many of your frames will be dark, and it is often impossible to find frame lines

between dark frames. Also, for negative film, insist that your processor print all the frames; tell them that these are pictures of stars and comets. Most processors will not print blank frames, and a casual look will tell a processor that your film is quite blank, even though you know otherwise.

Techniques

There are a number of things you can do to improve your comet pictures and make them uniquely your own, both before and after the comet arrives, and while it is here.

1. Practice. Halley's comet will be visible in the evening, night, and morning sky. Set up your camera and tripod before sunrise, before sunset, and in the middle of the night, some time during the year before the comet arrives. Use the film you intend to photograph the comet with, or experiment with several films to determine which one will work best for you. Make a series of exposures, both with your camera aimed at the horizon, and overhead, as you would for the comet. Keep a record of your exposure times. You will find that some exposures are obviously too dark, and some too light. From this test, you will also be able to determine just how much star trailing is acceptable to you.

2. Make a comet panorama. Comets are very large, but very dim. It takes several times more exposure to properly expose the dimmer parts of the tail than it does to properly expose the coma. In other words, if the coma is properly exposed the tail is underexposed; if the tail is properly exposed, the coma is overexposed. A panorama is a way to have some of your comet cake, and eat it, too.

The technique requires print film. Make two exposures of the comet, one to properly expose the coma, and a much longer one to properly expose the tail. Put your prints side-by-side, and match up the detail. You will find that, although the background density and the star trails will differ, you will have the most complete comet picture available to you with simple equipment. This technique may be the best use for your fast telephoto lens.

3. Make a comet family portrait. This is a good one to do once you know your basic exposure. Line up your shot of the comet, and then place your family or other loved ones in the frame in the foreground. Tell everyone to stay very still, and start your exposure. Near the beginning, flash a strobe toward them. (The people in the foreground must remain there during the entire time exposure for this to work.) You will have a record of your entire family, plus the comet—You Are There!

The difficult part of this trick is to determine how far away to stand with the strobe when you flash it. Remember, small strobes are designed to be used at the camera, to make a proper exposure at a small lens aperture. Since your lens will be set wide open, you will need to stand some distance behind your camera, holding the strobe to light your subject properly. Since different strobes have different light output there is no hard-and-fast rule that can be applied. Also, manual, automatic and dedicated strobes each have different characteristics.

In general, if you move your strobe back half as far behind your camera as your subject is in front of it, you reduce the light from the strobe by one f/stop. Moving back the same distance as your subject is in front reduces your light by two f/stops.

An easier solution to the problem might be simply to cover your strobe with a layer or two of white tissue paper or a white handkerchief, to reduce its light output while keeping it on your camera. Whichever method you choose should be practiced in advance. The best solution for how to use your strobe for this picture is to discuss it with someone at a reliable camera store near you.

With this simple equipment, and these simple techniques, you can get good comet pictures. Other techniques, such as telescope mounted cameras and clock-driven or hand-guided camera platforms, are used by serious amateur astronomers. If you are interested in using more sophisticated equipment and techniques, approach your local amateur astronomical

Edmund Halley: Historical Highlights

1656(?)	born London, England, 29 October
1673–76	Queen's College, Oxford
1676	moved to St. Helena Island (lat. 16° S) to catalog stars of the southern hemisphere
1677	M.A. Oxford University (by royal *mandamus*)
1678	published catalog of stars of the southern hemisphere; elected Fellow of the Royal Society of England
1682	married Mary Tooke (three children: Katherine and Margaret, 1688; Edmund, 1698)
1683	published paper on geology
1685–93	edited *Philosophical Transactions of the Royal Society*
1686	edited and published *Philosophia Naturalis Principia Mathematica* for Isaac Newton; published paper on trade winds and monsoons
1691	published paper on date and place of Julius Caesar's first landing in Britain
1692	formed company for salvaging wrecks using diving bell and helmet; published paper on geomagnetism
1693	appointed Deputy Comptroller of the Royal Mint
1698–1700	commissioned naval Captain in *HMS Paramore* to map magnetic variations
1701–03	published maps using isogonic (Halleyan) lines of equal magnetic variation
1704	appointed Savilan Professor of Geometry at Oxford
1710	published definitive version of Appolonius's *Conics*; awarded Degree of Doctor of Civil Laws; discovered proper motion of stars
1712	published Flamsteed's observations, *Historia Coelesti*
1715	published paper on stellar novae
1720	appointed Astronomer Royal of England
1729	elected to the Academie des Sciences, Paris
1743	died Greenwich, England, 14 January

Halley is most famous for his scheme for computing the motions of comets. He was particularly interested in the bright comet of 1682. Using the hypothesis that cometary paths are nearly parabolic, he considered the bright comets of 1531, 1607 and 1682 to be the same object, returning approximately every seventy-five years. He also concluded that this comet was seen in 1305, 1380 and 1456.

Halley announced that this comet should return in December of 1758. The comet was sighted by a Saxon peasant-farmer and self-taught amateur astronomer, Johann Palitzch, using a home-built seven-foot reflecting telescope, on December 25, 1758.

Halley also made notable advances in navigational astronomy. He determined a method for determining the distance from the Earth to the Sun by observing and timing transits of the Sun by Mercury and Venus.

society. They may have hand- or clock-driven camera mounts and telescopes for club member use. They may even have classes to instruct you in proper astronomical photographic techniques. If they do not have a multi-camera mount specifically for public use, you might be able to talk them into it as a public service that their club can provide so everyone interested can obtain long, guided exposures.

When all the hoopla is over approach your local planetarium, science museum, or art gallery, to see if they are interested in showing or exhibiting your local community's photographic efforts.

Whatever you do, remember, Halley's comet is a once-in-a-lifetime phenomenon. Use your camera, and enjoy.

For More Information

If you are interested in joining the International Halley Watch (IHW), request information from:

International Halley Watch
Jet Propulsion Laboratory
Mailstop T 1166 B3
4800 Oak Grove Drive
Pasadena, CA 91009

To purchase the IHW *Amateur Observer's Manual for Scientific Comet Studies* send to:

Sky Publishing Corp.
49 Bay State Rd.
Cambridge, MA 02238

or

Enslow Publications
Bloy St. and Ramsay Ave.
PO Box 777
Hillside, NJ 07205

Cost: $9.95

To obtain the IHW *Amateur Observer's Bulletin* write to:

Planetary Society
PO Box 91687
110 S. Euclid
Pasadena, CA 91109

Cost: free

Joe Laufer
Halley Comet Watch Newsletter
PO Box 188
Vincentown, NJ 08008

Quarterly publication. Cost: $4.00/year. Also sells *Comet Halley Bibliography.* Cost: $1.00/year

Dark Skies For Comet Halley Journal
c/o Astronomical League
Don Archer, Exec. Secretary

PO Box 12821
Tucson, AZ 85732

A quarterly journal. Cost: $4.00/year ($5.00 outside U. S. A.)

Comet News Service
PO Box TDR #92
Truckee, CA 95734

A quarterly newspaper. Cost: $5.00/year

Publishers of the *International Comet Quarterly* are:

Daniel Green
Smithsonian Astrophysical Observatory
60 Garden St.
Cambridge, MA 02138

Cost: $8.00/year
(Please do *not* use the title *International Comet Quarterly* in the mailing address.)

Astronomy Magazine
Astromedia Corp.
PO Box 97288
Milwaukee, WI 53202

Monthly. Cost: $21.00/year

Sky And Telescope Magazine
Sky Publishing Co.
49 Bay State Rd.
Cambridge, MA 02238

Monthly. Cost: $18.00/year

The publishers of the annual *Observer's Handbook* are:

Royal Astronomical Society of Canada
124 Merton St.
Toronto, Ontario
M4S 2Z2

Membership cost: $20.00/year

Publisher of the annual *Sky Almanac*, an astronomical calendar, is:

Guy Ottewell
Furman University
Greenville, SC 29613

Cost: $10.00

Edmund's Sky Guide and the *Magnitude 6 Star Atlas* are two good, usable star atlases. They are available from:

Edmund Scientific Co.
7785 Edscorp Bldg.
Barrington, NJ 08007

Cost: $5.00 and $11.00, respectively

For information concerning the Boothe Memorial Halley Comet Festival, March 22–23, 1986, contact:

Ron Przyblyo
Control of Operations
225 High Park Avenue
Stratford, CT 06497

Books

Ash, Russel, and Ian Grant: *Comets*, Bounty (1973)

Brandt, John C: *Comets: Readings from Scientific American*, W.H. Freeman (1981)

Brown, Peter L: *Comets, Meteorites, and Men*, Taplinger (1974)

Delsemme, A.H., ed.: *Comets, Asteroids, Meteorites*, University of Toledo (1977)

Donn, et al: *The Study of Comets* (NASA SP-393), U. S. Printing Office (1976)

Donn, Rahe, Wurm: *Atlas Of Cometary Forms*, (NASA SP-198) U. S. Printing Office (no date)

Seargent, David A: *Comets: Vagabonds of Space*, Doubleday (1982)

Wilkening, Laurel, ed.: *Comets*, University of Arizona (1982)

You will also find more information about comets, and about astronomy in general, by visiting your local planetarium and your public library.

Good observing.

Garry and Dwight